Contents

1 Chapter 1: Defining Excellence

Operational excellence isn't a destination; it's a continuous journey. It demands a clear understanding of what constitutes "excellent" within your specific context. This isn't about arbitrary standards; it's about defining success according to your organization's unique goals and values. Consider your mission statement, your long-term objectives, and the specific metrics that truly reflect your progress. Are you focused on maximizing efficiency, minimizing costs, or perhaps prioritizing customer satisfaction above all else? The answer will shape your definition of excellence and guide your subsequent actions. Ignoring this crucial first step, defining what excellence means to you, often leads to aimless striving and ultimately, disappointment.

Setting clear goals is paramount. These aren't simply broad aspirations; they require meticulous planning and articulation. Each goal should be Specific, Measurable, Achievable, Relevant, and Time-bound (SMART). Consider not just the final outcome but the incremental milestones along the way. Break down complex objectives into manageable tasks, assigning responsibility and deadlines to specific individuals or teams. Remember, clear goals provide direction, motivation, and a framework for measuring progress. Without them, your efforts will lack focus, resembling a ship sailing without a compass. Vague objectives will inevitably lead to vague results, making it impossible to accurately assess performance or identify areas for improvement. That precision—the clarity of your goals—is the bedrock of operational excellence.

Building high-performing teams is as crucial as defining those goals. It starts with careful recruitment, selecting individuals with the right skills and attitude. However, selecting the right team is only the beginning. You must foster an environment of collaboration and trust, where team members feel valued and empowered. Clearly define roles and responsibilities to avoid confusion and duplication of effort. Encourage open communication and feedback, creating a safe space for constructive criticism and idea-sharing. Regular team meetings, focused on progress and problem-solving, can significantly enhance performance. Think of it as cultivating a garden: You need the right seeds (personnel), fertile soil (environment), and consistent nurturing (communication and support) to reap a bountiful harvest. A team working cohesively, with clear roles and a shared understanding, is an invaluable asset in your pursuit of excellence. Conversely, a dysfunctional team, rife with conflict and lacking clear direction, will hinder even the most meticulously planned operations.

Effective communication is the lifeblood of any successful project, particularly in pursuit of operational excellence. It's about more than just transmitting information; it's about ensuring a shared understanding among all team members, stakeholders, and relevant parties. Transparency is key. Keep everyone informed of progress, challenges, and changes in plans. Utilize various communication channels—meetings, emails, project management software—to suit different needs and preferences. Don't underestimate the power of face-to-face interactions for building relationships and fostering trust, especially when dealing with potentially sensitive or complex issues. Lack of communication leads to misunderstandings, missed deadlines, and ultimately, compromised results. Open, honest, and consistent communication should be a cornerstone of your approach to excellence. This ensures everyone is on the same page, contributing effectively toward the shared goals, and minimizes costly errors or delays caused by miscommunication.

1.1 Setting Clear Goals

Effective goal-setting isn't merely about writing down aspirations; it's about crafting a roadmap to success. Begin by defining your overarching objective. What is the ultimate aim of your project? Be specific; avoid vague terms like "improve efficiency." Instead, quantify your goals. For instance, aim to "reduce production time by 15% within the next quarter" or "increase customer satisfaction scores by 10 points." These concrete targets provide measurable benchmarks for your progress.

Next, break down that large objective into smaller, manageable tasks. Think of it as creating a hierarchical structure. Your primary goal sits at the top, branching out into secondary objectives, then tertiary tasks, and so on. This hierarchical breakdown simplifies the process, making the large, daunting goal feel more achievable. Each smaller task should also have its own clear deadline and assigned responsibility. This eliminates confusion and ensures accountability.

Remember the SMART framework: Specific, Measurable, Achievable, Relevant, and Time-bound. This simple acronym ensures your goals are well-defined and practical. An achievable goal is challenging yet realistic, keeping you motivated without discouraging you. Relevance ties your goals directly to your overall project objectives, preventing wasted effort on irrelevant tasks. Finally, a time-bound goal provides a sense of urgency and helps prevent procrastination.

Collaboration is crucial. Involve your team in the goal-setting process. Their input will not only enhance the quality of the goals, but it will also increase their buy-in and commitment. Open communication is paramount. Regularly review your progress, celebrating milestones along the way and adapting your strategy as needed. Unexpected hurdles will inevitably arise; flexibility is key to navigating them effectively. Regular monitoring is not just about tracking progress; it's about identifying potential roadblocks early on. Consistent tracking allows for prompt adjustments, preventing minor issues from escalating into major problems. Use a variety of tools; spreadsheets, project management software, even simple checklists can all be useful. The important thing is to choose a method that suits your team and project. Remember to incorporate feedback loops into your monitoring system.

Consider using a visual representation of your goals, like a Gantt chart or a Kanban board. These visual tools provide a clear overview of your progress and highlight dependencies between tasks. They also facilitate communication and help you identify potential conflicts or delays. Regular check-ins with your team will ensure everyone is on the same page and working towards the common goal efficiently. This not only boosts team morale, but it also ensures everyone is aligned.

Finally, remember that goal-setting is an iterative process, not a one-time event. Regularly review and refine your goals based on your progress and changing circumstances. This adaptability is essential for navigating unforeseen challenges and ensuring you remain on track towards your ultimate objective. Don't be afraid to adjust your strategy as you learn and grow throughout the project lifecycle. Adaptability is crucial for long-term success. Continuously assessing your goals and strategies ensures that your project remains aligned with its original purpose and adapts to the ever-changing environment. Remember, the ability to adjust and refine your approach is a critical skill that will lead to enduring success.

1.2 Building Strong Teams

Teamwork isn't merely about individuals working together; it's about creating a synergistic entity, a force greater than the sum of its parts. Building a strong team necessitates more than just hiring skilled people; it requires cultivating a shared vision, fostering open communication, and nurturing a culture of mutual respect and trust. Success hinges on these intangible yet crucial elements.

Effective team building starts long before the first project meeting. The selection process itself is critical. Look beyond technical skills; assess individuals' personalities, work styles, and interpersonal abilities. A team composed of individuals with complementary strengths and diverse perspectives is inherently more resilient and innovative. Consider using personality assessments or behavioral interviews to gain deeper insights into candidate suitability. Remember that a cohesive team can overcome skill gaps, while a team lacking cohesion will struggle regardless of individual brilliance.

Establishing clear roles and responsibilities is paramount. Ambiguity breeds conflict and inefficiency. Each team member should understand their specific contributions, their accountabilities, and how their work integrates with the larger project goals. This clarity prevents duplication of effort, ensures accountability, and promotes a sense of ownership. A well-defined structure fosters individual responsibility whilst ensuring smooth collaborative working. Regular check-ins and open discussions can help to ensure that roles remain relevant and that responsibilities are fairly distributed as the project evolves.

Open communication isn't simply about regular meetings; it's about creating an environment where everyone feels comfortable sharing ideas, concerns, and feedback. Encourage open dialogue, active listening, and respectful debate. Establish clear communication channels and utilize various methods depending on the urgency and nature of the information. This might range from quick instant messages for immediate updates to formal meetings for strategic discussions or in-depth project reviews. Remember that effective communication isn't just about transmitting information; it's about ensuring that information is received, understood, and acted upon.

Conflict is inevitable in any team. However, how you manage conflict is critical to a team's success. Establish a structured approach to conflict resolution, emphasizing respectful communication and collaborative problem-solving. Encourage team members to express their disagreements openly but constructively. Facilitate discussions that focus on finding solutions rather than assigning blame. Remember, well-managed conflict can lead to increased creativity and stronger team bonds.

Beyond project-specific tasks, foster team cohesion through informal activities. Organize team-building exercises, social events, or even casual lunches to build camaraderie and strengthen relationships. These activities create opportunities for individuals to connect on a personal level, fostering trust and mutual understanding. Remember that a strong team is built on personal connections as well as shared professional goals. Team morale is often the key to success in the most demanding circumstances.

Regular feedback is essential for both individual and team growth. Conduct regular performance reviews that provide both constructive criticism and positive reinforcement. Encourage peer feedback to promote a culture of continuous learning and improvement. This also ensures that team members feel valued and appreciated for their contributions. It also allows you to make proactive adjustments to support those who might be falling behind.

Finally, celebrate successes both big and small. Acknowledge and reward individual and team achievements to reinforce positive behaviours and motivate ongoing performance. This could range from simple verbal praise to more formal recognition programs. Remember, positive reinforcement significantly enhances team morale and encourages future success. The feeling of accomplishment is what keeps a team motivated and energised to conquer larger goals together. A team who feels supported and celebrated will always work better together.

2 Chapter 2: Process Mapping

Process mapping isn't just drawing pretty pictures; it's about understanding the lifeblood of your operations. Think of it as an X-ray, revealing hidden inefficiencies and bottlenecks that drain resources and stifle productivity. You'll begin by meticulously documenting each step of a chosen process, from start to finish. Consider using simple, universally understood symbols—swim lane diagrams are particularly useful for visualizing the flow of work across different departments or individuals.

Detailed descriptions are vital. Don't just list actions; explain why each step is necessary, its duration, and the resources it consumes. Identify handoffs between different team members or departments, noting any potential delays or communication breakdowns. The goal is to build a comprehensive, accurate representation of the current state. Only then can you effectively identify areas for improvement.

Once you have a clear picture, start looking for bottlenecks. These are points where work slows down or gets stuck. Perhaps a particular step is overly time-consuming, or maybe there's a lack of clarity in responsibilities leading to delays. Analyze each step critically; are there redundant steps that can be eliminated? Could tasks be consolidated or automated?

Optimization goes beyond simply removing steps. It's about intelligent resource allocation. Are you effectively utilizing your personnel, equipment, and materials? Are there instances of underutilization or unnecessary duplication? By systematically evaluating resource use, you can pinpoint areas where adjustments can boost efficiency and reduce waste. The process map becomes your roadmap for strategic resource allocation, aligning personnel and materials with the most critical tasks.

Consider using technology to enhance the process mapping exercise. There are software solutions available that automate many of the tasks, from creating visual diagrams to analyzing data for bottlenecks. Remember, the goal is not just to create a visual representation, but to improve efficiency. This process doesn't just benefit your immediate project, it lays a foundation for ongoing improvement in overall operational efficiency and reduces long-term costs.

Remember, process mapping is an iterative process. Your initial map is a starting point, not the final product. After implementing changes based on your initial analysis, revisit and update the map. Continuous monitoring allows you to track the impact of your improvements and identify new areas for optimization. Don't be afraid to experiment; some strategies may yield better results than others. The key is to remain data-driven, tracking key metrics to ensure your efforts are paying off.

Finally, communicate your findings clearly. Share your process maps and analysis with your team, emphasizing areas of improvement and the rationale behind proposed changes. A transparent and collaborative approach is key to successfully implementing changes and gaining buy-in from all stakeholders. Engage your team in the process; their experience and perspectives are invaluable. By fostering collaboration and a shared understanding of the revised process, you lay the groundwork for sustainable efficiency improvements. Remember to document all changes and their associated impact, creating a repository of knowledge that you can refer to for future process improvement initiatives. This iterative approach ensures continuous refinement and adaptation, making your operational processes more resilient and adaptable to change over time. This forms a vital link to subsequent chapters addressing data-driven decisions and continuous improvement strategies.

2.1 Identifying Bottlenecks

Identifying bottlenecks is crucial for optimizing any operational process. Think of it like a river: a narrow section will restrict the flow, no matter how wide the river is elsewhere. Similarly, in your workflow, even a small, seemingly insignificant step can significantly impact overall efficiency. Finding these "narrow sections" is the key to unlocking significant improvements.

Effective bottleneck identification requires a multifaceted approach. You can't rely solely on intuition; you need concrete data. Start by carefully mapping your processes. Use a flowchart, a swim-lane diagram, or any visual tool that clearly shows the steps involved, the handoffs between departments or individuals, and the time taken for each stage. This visual representation makes it much easier to spot potential areas of congestion.

Once you have your process map, pay close attention to the points where work piles up. Are there certain tasks that take disproportionately long? Are there specific individuals or teams consistently overloaded? These are prime candidates for bottlenecks. Quantify these delays. Use time studies, tracking software, or simply ask your team members about their daily experiences. Objective data allows for a more rational analysis compared to subjective estimations.

Beyond individual tasks, consider broader system-level factors. Are there insufficient resources, such as equipment, software, or personnel, hindering progress? Are there approval processes that create unnecessary delays? Inadequate training or unclear instructions can also trigger bottlenecks. Examining your resource allocation strategy is also vital. Are resources being distributed effectively or are some areas starved while others are overloaded? This uneven distribution can drastically slow down the entire operation.

Let's use a real-world example. Imagine a manufacturing plant where the packaging process is slower than the production line. This creates a bottleneck: finished products pile up, slowing down production and impacting delivery times. Identifying this bottleneck allows for strategic intervention; perhaps investing in faster packaging equipment, hiring additional personnel, or streamlining the packaging process itself. The solution will depend on a thorough understanding of the root cause.

The analysis shouldn't stop at identification. Once you've pinpointed the bottlenecks, prioritize them based on their impact on overall efficiency. Use metrics like cycle time, throughput, and defect rates to gauge their severity. Tackling the most impactful bottlenecks first will deliver the quickest and most significant improvements. Remember that fixing one bottleneck can sometimes reveal others hidden previously.

This iterative process is critical. After implementing solutions, regularly monitor the impact of your changes. Are they effective? Has a new bottleneck emerged? Continuous monitoring and evaluation are vital for sustained operational excellence. Don't be afraid to adjust your strategies, as what works today might not work tomorrow.

Ultimately, identifying bottlenecks is not just about finding problems; it's about driving continuous improvement. It's about cultivating a culture of observation, data-driven decision making, and a willingness to adapt and refine processes over time. By embracing this iterative approach, you'll not only improve efficiency but also foster a more proactive and resilient organization. This mindset of continuous improvement is key to achieving sustainable operational excellence.

Remember that the key to successfully identifying and resolving bottlenecks lies in combining rigorous data analysis with a deep understanding of the human element involved in your processes. Don't underestimate the value of direct feedback from your team. They are often the first to see the issues and offer creative solutions.

Effective communication is paramount throughout this entire process. Keep your team informed about identified bottlenecks, planned solutions, and the progress being made. Transparency fosters trust and buy-in, critical elements in successfully implementing changes. Ignoring your team and pushing changes unilaterally can often create additional challenges and undermine the entire improvement process.

Moreover, consider the potential impact of technology. Automation and process optimization software can streamline workflows, potentially eliminating bottlenecks or reducing their impact. Before investing in any technology, however, ensure that it directly addresses a specific bottleneck and aligns with your overall strategy. Avoid adopting technology just for the sake of it; focus on solutions that will genuinely improve efficiency.

Finally, remember that identifying bottlenecks is an ongoing process. As your organization evolves and adapts to new challenges and opportunities, your processes will inevitably change, bringing with them the potential for new bottlenecks to emerge. Develop a system for regularly reviewing and reassessing your workflows to proactively identify and address these emerging challenges before they significantly impact your operational excellence. This proactive approach is crucial for maintaining peak performance and achieving long-term success.

2.2 Streamlining Workflows

Identifying bottlenecks is the first step, but understanding why those bottlenecks exist is crucial for effective workflow streamlining. A simple time study might reveal a slow process, but delving deeper reveals the root cause - perhaps inadequate training, outdated equipment, or a poorly designed process itself. Addressing the underlying issue, not just the symptom, is key to lasting improvement. This requires careful observation, thoughtful analysis, and a willingness to challenge established practices. Don't just fix the immediate problem; eliminate the conditions that caused it in the first place. This proactive approach ensures future efficiency gains.

Effective workflow streamlining isn't just about speed; it's about optimizing the entire process. It's about finding the right balance between efficiency and quality. Cutting corners might seem like a quick win, but it often comes at the cost of errors, rework, and ultimately, diminished quality of output. A streamlined process should be both fast and reliable, producing consistent, high-quality results. This balance is achieved through careful planning, meticulous execution, and constant monitoring of performance indicators.

Prioritize tasks based on their value and urgency. Not all tasks are created equal; some contribute significantly more to overall project goals than others. Focus your efforts on high-impact activities first, delegating or automating lower-priority tasks where possible. Implementing a prioritization system, such as the Eisenhower Matrix, can help you systematically manage your workload, ensuring that your time and resources are allocated effectively. This prevents you from getting bogged down in less important tasks and allows you to focus on those which truly move the needle.

Automation plays a critical role in streamlining workflows. By automating repetitive, manual tasks, you free up valuable time and resources for more strategic initiatives. However, automation is not a one-size-fits-all solution. Carefully evaluate which tasks are suitable for automation and ensure that the automation system integrates seamlessly with your existing workflows. Poorly implemented automation can actually increase inefficiencies and create new problems, so careful planning and thorough testing are essential before full-scale implementation.

Technological advancements offer numerous opportunities for process improvement. From project management software that centralizes communication and task assignments to specialized tools designed for specific workflows, technology can significantly enhance efficiency. The key lies in selecting the right technology to meet your specific needs and ensuring that your team is properly trained on how to use it effectively. Remember that technology is only as good as the people using it. Invest in comprehensive training and ongoing support to maximize the benefits of your technological investments.

Collaboration is paramount in streamlining workflows. Effective communication, both within and across teams, is essential for identifying bottlenecks, sharing ideas for improvement, and coordinating efforts. Establishing clear communication channels, regular meetings, and a shared platform for documentation fosters collaboration and reduces the risk of misunderstandings and delays. Consider implementing project management software or communication tools to support and enhance collaboration, ensuring that everyone is on the same page and working towards the same goals.

Regular review and adjustment are critical to maintaining a streamlined workflow. What works well today might not be optimal tomorrow. As circumstances change, so too should your processes. Implement a system for regularly monitoring key performance indicators (KPIs) and conducting process audits. Use the data gathered to identify areas for

improvement, and don't hesitate to make adjustments as needed. This proactive approach ensures that your workflows remain efficient and effective over time, adapting to evolving needs and external factors.
Finally, remember that streamlining workflows is an ongoing process, not a one-time fix. Continuous improvement is key. Encourage feedback from your team, actively seek out new ideas, and embrace a culture of innovation. The goal is not just to achieve efficiency today, but to build a system that adapts and improves continuously, driving sustained operational excellence into the future. By embracing these principles, you can transform your workflows from bottlenecks to highly efficient engines driving peak performance. The payoff is not just greater efficiency, but also increased employee satisfaction and a stronger competitive advantage.

2.3 Optimizing Resources

Resource optimization isn't about squeezing every last drop; it's about strategic allocation for maximum impact. Think of it like a skilled chef—they don't hoard ingredients, they use each one precisely, enhancing the overall dish. Similarly, understanding where your resources are currently deployed is crucial before you even think about optimizing them. A thorough assessment, perhaps utilizing a resource matrix, will reveal imbalances and hidden inefficiencies. This step is fundamental; without it, optimization efforts become mere guesswork.
Identifying bottlenecks is vital. These aren't just slow points; they're the chokepoints limiting your overall throughput. Imagine a highway with a single lane bottleneck; regardless of the speed elsewhere, the overall traffic flow suffers. Similarly, in your operation, pinpoint the processes, resources, or teams consistently delaying progress. These are prime targets for immediate attention. Sometimes, a simple ad-

justment—a better workflow, additional staffing during peak hours, or improved technology—can dramatically improve efficiency.

Consider automation. Repetitive tasks, prone to human error, are ripe for technological solutions. Automation doesn't necessarily mean robots; even simple software can streamline processes, freeing up valuable human resources for more strategic tasks. Think of data entry, report generation, or scheduling—tasks easily automated, allowing your team to focus on higher-level analysis and problem-solving. Remember, technology should augment, not replace, human expertise. It's about a collaborative effort, a partnership between human ingenuity and technological efficiency.

Delegation and empowerment go hand in hand. Micromanagement is the enemy of efficiency; it creates bottlenecks and stifles innovation. Empower your team to take ownership of their tasks, offering support and guidance but allowing them the autonomy to find solutions. This builds trust and fosters a sense of responsibility, leading to greater efficiency and engagement. A clear delegation strategy, coupled with regular check-ins and feedback, ensures everyone is working effectively and contributing to the overall optimization strategy.

Don't underestimate the power of training and development. Investing in your team's skills is an investment in your operation's efficiency. Regular training can improve proficiency, reduce errors, and enhance problem-solving abilities. Consider cross-training as well—this expands individual capabilities and creates a more flexible workforce, capable of handling a wider range of tasks and adapting to changing demands. The payoff from a well-trained, confident workforce far outweighs the initial investment.

Regular review and refinement are crucial. Optimization isn't a one-time fix; it's an ongoing process. Set up a system for monitoring key performance indicators (KPIs) and regularly reviewing your resource allocation. Are your changes working as intended? Are new bottlenecks emerging? Are there areas for further improvement? Continuous monitoring and adaptation are key to maintaining optimal resource utilization and ensuring sustained operational excellence. This cyclical process ensures your optimization strategy remains dynamic and responsive to changing conditions.

Remember, optimization is a holistic endeavor. It requires analyzing your entire system—people, processes, technology—to identify areas for improvement. It isn't simply about cutting costs; it's about maximizing value and efficiency across the board. A well-optimized system will exhibit enhanced productivity, reduced waste, and improved overall performance. The key is to approach this systematically, using data to guide your decisions and focusing on continuous improvement. This requires dedication and effort, but the returns—improved productivity, a more engaged team, and a more profitable operation—make it a worthwhile investment.

3 Chapter 3: Data-Driven Decisions

Gathering the right data is paramount. Don't just collect information; strategically select key performance indicators (KPIs) that directly reflect your goals. For example, if your aim is to reduce production time, track metrics like cycle time, throughput, and defect rates. Focus on the data that truly matters, avoiding the trap of data overload that obscures the crucial insights. Remember, precision trumps volume. Carefully consider which metrics will provide the most accurate reflection of your progress toward achieving operational excellence. This careful selection will be crucial for efficient analysis and the development of effective strategies for improvement. Selecting relevant KPIs forms the groundwork for data-driven decision-making in the context of overall operational excellence. Choose wisely, and the rest will follow.

Analyzing your data involves more than just looking at numbers. It requires pattern recognition and identification of correlations. You might notice, for instance, that a specific production step consistently causes delays, or perhaps customer satisfaction dips after a product update. Use statistical tools and data visualization techniques—charts, graphs, dashboards—to unveil these patterns. This process transcends mere observation; it involves critical thinking, drawing inferences from the raw data, and transforming it into actionable intelligence. Tools like regression analysis or even simple scatter plots can reveal surprising correlations, providing clues for more impactful interventions. This understanding allows for the formulation of solutions more effectively. The insights gleaned from data analysis are not merely theoretical; they should directly inform your decisions. Consider a scenario where your analysis shows that employee training significantly reduces error rates. This insight directly justifies investing more in training programs. Or, if your sales data reveals a sudden drop in a particular product line, you may need to adjust marketing strategies or redesign the product. This isn't about making changes randomly; it's about tailoring your actions to concrete evidence, maximizing the impact and ensuring efficient resource allocation. The essence lies in translating raw data into concrete steps to enhance efficiency and performance. The power of data lies not in its accumulation, but its transformation into action.

Predictive analytics takes data analysis a step further. It's not just about understanding the past; it's about forecasting future trends. This enables proactive measures. Analyzing historical sales data can help predict future demand, allowing for optimized inventory management. Similarly, analyzing machine maintenance records can predict potential equipment failures, allowing for timely repairs and preventing costly downtime. This proactive approach minimizes disruption and maximizes resource utilization. It's essentially the difference between reacting to problems and preventing them entirely.

Effective reporting is essential for communicating your findings. The purpose is clarity and impact. Use clear, concise language; avoid overwhelming readers with technical jargon. Choose the right visualization tools—charts, graphs, maps—to showcase your findings in an easily digestible format. Presenting your findings visually and clearly allows for quicker comprehension and facilitates easier adoption of recommended changes. Consider your audience, tailor your report to their understanding, and make sure your recommendations are clear, concise, and actionable. The goal is not just to present data, but to inform decision-making and inspire change. This ensures your insights will lead to tangible improvements in operational excellence.

3.1 Collecting Key Metrics

Choosing the right metrics is crucial. You wouldn't navigate by looking at only the rearview mirror, would you? Similarly, focusing on irrelevant data hinders your progress. Effective metric selection requires a deep understanding of your operational goals and the processes that drive them. Start by identifying the key areas of your operation, and within those areas, pinpoint specific processes contributing to (or detracting from) your success. Consider your organization's overall objectives - are you focused on growth, efficiency, customer satisfaction, or something else? This overarching strategy will define the types of metrics most relevant to your progress. Remember, fewer, well-chosen metrics are better than a flood of meaningless numbers. Prioritize a handful of key performance indicators (KPIs) that accurately reflect your progress toward achieving your defined goals.

The process of choosing and implementing these crucial metrics is iterative. It's rarely a case of 'set it and forget it.' Expect to refine your metrics regularly. As your project evolves, you might discover that what was once critical is now less relevant, or that you've missed a vital indicator. This ongoing review is key to maintaining your focus and ensuring your metrics continue to provide valuable insights. Don't hesitate to adjust your collection and analysis methods as needed - the goal is to gain an accurate picture of your progress, so the process itself should be dynamic. This is not just about picking the right metrics, but also about understanding when to change them. The context of the situation must be taken into consideration, and you should avoid being too rigid in your approach. Change is inevitable, and being adaptable in your method is crucial to being successful.

Data, by itself, is useless. It needs context to transform into actionable insights. Consider how each metric relates to your overall objectives. What story does the data tell you? For example, if your goal is to reduce production lead times, metrics such as the average production cycle time and the number of production bottlenecks would be vital. Analyzing these metrics over time provides a clear picture of whether your interventions are effective. Moreover, remember to always correlate your data. Is a decrease in one metric associated with an increase in another? These relationships often provide valuable insights into the underlying processes. It's often the unexpected correlations that offer the greatest potential for improvement. Do not just look at individual metrics in isolation. Analyzing this data holistically will paint a more complete picture and provide a more well-rounded view of performance.

Effective data collection requires careful planning. Establish a clear system for collecting data from the beginning. Will you use manual methods, spreadsheets, dedicated software, or a combination thereof? Consistency is paramount; ensuring data accuracy is fundamental to making informed decisions. Consider what technology is best suited to your needs, and make sure the system is easy to maintain and use by your team, so they can contribute to a holistic database. Without an easy to follow system, the data collecting process itself can become a bottleneck, thus hindering your whole project. Ensure that everyone in the team understands the importance of timely and accurate data entry. If the system is complex, or the entry procedure is cumbersome, this will almost certainly lead to inaccuracies, and potentially, the whole system can be deemed useless.

Beyond the technicalities of data collection, consider the human element. How will you motivate your team to collect and input this data correctly? Effective communication is key here. Explain the importance of accurate data for the project's success. Make sure to highlight the benefits of doing this to show them that the data collection process is valuable. Show them how their input contributes to the overall goals, and how it makes a positive impact on the operation. Make the process as easy and user-friendly as possible. The more user-friendly it is, the more likely they are to comply. Involve your team in the process of developing the system if you can. This helps foster a sense of ownership and makes them more inclined to use the system accurately. You need buy-in from your team for your metric system to succeed.

Finally, don't forget the importance of visualization. Raw data can be overwhelming. Transforming it into clear, concise visuals, such as graphs and charts, significantly improves comprehension and communication. Use these visualizations to present your findings to stakeholders, highlight key trends, and support your improvement initiatives. Simple, well-designed visuals make your insights readily accessible to everyone involved and contribute to a shared understanding of progress. This facilitates communication and ensures that everyone is on the same page. Visualisations make it easier to see trends, spot outliers, and communicate the implications of the data clearly and concisely, allowing even those not directly involved in the data analysis process to understand its significance. Effective data visualization is a critical part of a complete data-driven approach.

3.2 Analyzing Performance Data

Analyzing performance data isn't just about crunching numbers; it's about uncovering hidden stories within your operational landscape. Think of it as detective work, where the clues are scattered across spreadsheets and databases, revealing patterns of success and areas ripe for improvement. You need to be curious, patient, and persistent in your pursuit of understanding. The right metrics, meticulously collected, act as your bread crumbs, guiding you to the heart of operational efficiency.

Understanding your data requires more than just technical skill; it demands a clear understanding of your organization's goals and processes. You must know what you are looking for before embarking on your analysis. Before diving into complex statistical models, ensure you've identified the key performance indicators (KPIs) that directly reflect your operational objectives. Are you aiming to reduce production time? Improve customer satisfaction? Increase profitability? These overarching goals dictate the specific metrics you'll track and analyze, guiding your data selection process. The wrong metrics won't reveal the whole story.

Once you have your data, choosing the right analytical tools is essential. Depending on the complexity of your data and your analytical skills, you might employ simple spreadsheet software, statistical packages, or advanced business intelligence (BI) tools. Each tool offers a unique set of capabilities; the key is to choose the one that best suits your needs and level of expertise. Don't feel pressured to use the most sophisticated tool; often, a simpler approach yields perfectly satisfactory results, especially during the initial stages of data exploration. Remember, the goal is insight, not the tool itself.

Now, let's discuss the actual analysis. Begin with descriptive statistics:

calculate averages, medians, standard deviations. These fundamental calculations give you a general overview of your data, revealing central tendencies and variability. Visualizations, such as histograms, bar charts, and scatter plots, are incredibly powerful tools for interpreting these statistics. They help you quickly identify trends, patterns, and outliers, often revealing insights hidden within raw numbers. Don't underestimate the power of visual representation; a well-constructed chart can tell a thousand words.

Beyond descriptive statistics, delve into more advanced analytical techniques like regression analysis or time series analysis, depending on your data type and research questions. Regression analysis helps uncover relationships between variables. For instance, you might analyze the relationship between employee training hours and product defect rates. Time series analysis is particularly useful for identifying trends and seasonality in your data, allowing you to forecast future performance and make proactive adjustments. Remember that the choice of technique must align with the type of data and questions you are trying to answer.

Interpretation is as crucial as analysis. Once you've run your analyses, don't simply report the numbers; tell a story with your data. Identify meaningful patterns, draw conclusions, and relate your findings back to your initial objectives. What trends emerged? What factors influenced performance? Where are the opportunities for improvement? Your interpretation should provide actionable insights that directly inform operational decisions. Remember, the goal is not just to analyze the data, but to use that analysis to improve the way you work.

Finally, remember to document your entire process. This is often overlooked, but detailed documentation is vital for transparency, reproducibility, and future reference. Clearly describe your data sources, your analytical methods, your findings, and your conclusions. This documentation will be invaluable for future analyses and will help others understand your work. Maintain a well-structured report, complete with clear visualizations. This not only presents your findings effectively, but also acts as a testament to your diligent work.

3.3 Using Insights for Improvement

Analyzing your data is only half the battle; understanding it and leveraging those insights for concrete improvement is where true impact lies. Don't just collect numbers; unearth the stories hidden within them. What are your key performance indicators (KPIs) really telling you? Is a seemingly minor fluctuation in customer satisfaction hinting at a larger underlying issue, perhaps a process breakdown or a change in customer expectations?

A significant drop in productivity in a specific department might not be due to employee underperformance but instead, a lack of necessary resources or outdated equipment. Perhaps an increase in production errors points to a need for improved training or a reevaluation of the workflow itself. The devil, as they say, is in the details. Dig deep; question assumptions. This isn't just about identifying problems; it's about developing a genuine understanding of their root causes.

This is where your process maps from Chapter 2 become invaluable. Overlay your data analysis onto those visual representations of your workflows. Where are the bottlenecks appearing in relation to your key metrics? Are certain stages consistently underperforming or generating significant errors? This visual connection helps to contextualize the raw data, turning abstract numbers into actionable insights that you can clearly demonstrate.

Remember, data analysis isn't a one-time event; it's an iterative process. Continuously monitor your KPIs, and be prepared to adjust your strategies. What worked last quarter might not be effective next quarter. The business world is dynamic; your approach to optimization needs to be equally flexible and responsive. Regularly revisit your data, seeking fresh perspectives and new opportunities for improvement.

Once you've identified areas needing attention, prioritize improvements based on their potential impact and feasibility. Start with quick wins—relatively easy changes that deliver significant results. These early successes build momentum and demonstrate the value of a data-driven approach, motivating your team to embrace further improvements. Small victories can have a tremendous impact on morale.

But avoid getting bogged down in small details. Maintain a focus on the overall picture, constantly checking if your improvements align with your overarching goals defined in Chapter 1. Each change should contribute to the larger objective; otherwise, you risk wasting resources on improvements that lack strategic significance. Prioritization is key.

Implementing changes effectively requires careful planning and communication. Don't just spring new procedures on your team; involve them in the process. Explain the reasoning behind the changes, address their concerns, and provide the necessary training and support. Remember, change management is as crucial as the changes themselves. Employee buy-in is essential for successful implementation.

The success of any change initiative depends on how well it's implemented, monitored, and improved over time. Therefore, establish clear metrics to track the impact of your changes and regularly review these results. This continuous feedback loop is essential for identifying unintended consequences and fine-tuning your approach. Never assume that a change, once implemented, will remain effective indefinitely.

Furthermore, create a system for documenting your successes and failures. This creates a knowledge base that will inform future improvements and prevent you from repeating mistakes. This system doesn't

only help you but also provides a valuable resource for future team members. Knowledge sharing is crucial for sustained operational excellence. The goal isn't just to improve but to systematically improve and institutionalize those improvements.
This chapter emphasizes the importance of combining data-driven analysis with a deep understanding of your operational processes and your team. It's not just about numbers; it's about understanding what those numbers mean within the context of your organization's operations and the larger strategic goals. Use the tools and techniques discussed in the previous chapters to unlock the potential of your data and steer your project towards unprecedented levels of operational excellence. The journey to peak performance is ongoing; continuous improvement is the key to sustained triumph.

3.4 Predictive Analytics

Predictive analytics leverages historical data to forecast future outcomes, a powerful tool for proactive decision-making in operational excellence. It moves beyond simply understanding past performance; instead, it allows for anticipating potential challenges and opportunities, enabling strategic adjustments before they significantly impact operations. Imagine, for instance, predicting equipment failure rates, allowing for preventative maintenance scheduling and avoiding costly downtime. This is not merely reactive problem-solving, it's proactive operational management.
Effective implementation requires a robust data infrastructure, capable of collecting, storing, and processing large volumes of relevant information from various sources. This necessitates careful selection of key performance indicators (KPIs), ensuring they accurately reflect operational efficiency and align with overall strategic goals. Poor data quality

will inevitably lead to inaccurate predictions, undermining the entire process. Therefore, meticulous data cleansing and validation are essential initial steps. Without clean and reliable data, even the most sophisticated algorithms will yield misleading results.

The choice of predictive modeling technique depends heavily on the specific operational context and the nature of the data. Simple time series analysis might suffice for forecasting predictable trends, such as seasonal sales fluctuations. However, for more complex scenarios involving multiple interacting variables, more advanced techniques like regression analysis or machine learning algorithms become necessary. Remember, the goal is not simply to use the most complex method available, but rather to select the approach best suited to the problem at hand, balancing predictive accuracy with model interpretability and ease of implementation. Consider the resources available, both in terms of personnel expertise and computational power.

Interpreting the model's output is crucial. A predictive model is only as useful as our understanding of its predictions. Therefore, clear visualization techniques, such as dashboards and interactive reports, are critical. These tools enable stakeholders at various levels to grasp the key findings and act upon them effectively. The ability to communicate complex analytical insights in a straightforward, understandable manner is vital for successful implementation of predictive analytics within an organization. Remember, a sophisticated model that remains obscure is ultimately ineffective.

Finally, successful implementation of predictive analytics requires an iterative process of continuous improvement. Regularly review the model's performance against actual outcomes, identifying areas for refinement or recalibration. The real world is rarely perfectly predictable, and external factors can influence operational performance in unforeseen ways. Adapting the predictive model to account for these unforeseen changes is key to maintaining its accuracy and relevance over time. This continuous feedback loop transforms predictive analytics from a one-time project into a dynamic, adaptive component of a continuously improving operational system. Treat it as an evolving process rather than a static solution. Regular audits and updates are essential. Consider assigning a dedicated team to monitor performance and adjust parameters as needed.

The benefits of predictive analytics extend far beyond simple forecasting. By identifying potential bottlenecks and risks ahead of time, organizations can proactively allocate resources, optimize processes, and ultimately enhance overall operational efficiency. Early detection of problems allows for timely intervention, minimizing disruptions and maximizing output. It empowers informed, data-driven decision-making at all levels, leading to a more robust and resilient operational system. Consider this an investment in long-term operational success, not merely an expenditure on a specialized tool. By integrating this approach into the overall operational strategy, organizations can foster a culture of proactivity and optimization.

3.5 Reporting & Visualization

Effective reporting and visualization are not merely about presenting data; they're about transforming raw numbers into compelling narratives that drive action. Imagine presenting a dense spreadsheet of figures to your executive team; they'll likely glaze over. Instead, picture a concise, visually engaging dashboard highlighting key performance indicators (KPIs) with clear trends and actionable insights. That's the power of effective reporting and visualization. It's the bridge between complex data analysis and strategic decision-making.

Creating impactful visualizations requires understanding your audience.

What are their priorities? What information is most relevant to them? A chart suitable for a technical team might overwhelm a senior manager who needs a high-level summary. Therefore, tailor your visualizations to the specific needs and understanding of the individuals who will be consuming the information. This isn't about dumbing down the data; it's about clarity and focus. Prioritize the most critical metrics, and make it easy to see patterns and anomalies.

Choosing the right chart type is paramount. A simple bar chart might suffice for comparing performance across different departments. A line graph can reveal trends over time, while a scatter plot might identify correlations between variables. Don't feel pressured to use overly complex charts; often, simplicity wins the day. A well-designed pie chart effectively shows the proportional distribution of a whole. However, avoid using too many pie charts in a single presentation; they can become overwhelming quickly, and they aren't effective for showing complex changes over time.

Beyond the chart itself, the supporting narrative is critical. Numbers don't speak for themselves. Your report needs a clear, concise introduction summarizing the key findings. Each chart or graph should have a brief caption explaining what it shows and its implications. A well-written narrative turns data points into a story, making it easier for the audience to understand and engage with your findings. Never underestimate the power of a good title - it should concisely encapsulate the report's essence.

Effective reporting goes beyond simply presenting findings; it also involves anticipating questions. Consider potential objections or areas of uncertainty your audience might have and address them proactively in your report. This shows forethought and strengthens your credibility. Remember, the goal isn't to just present data, but to initiate productive discussion and drive action toward operational excellence. Provide recommendations for improvement based on your analysis; don't just state the problems, offer solutions.

For instance, if your analysis reveals a bottleneck in a specific process, clearly identify the bottleneck, and then suggest specific strategies to address it - perhaps by automating a manual task, re-allocating resources, or improving employee training. This proactive approach transforms your report from a passive presentation into a tool for organizational change. Such tangible action steps set your reporting apart from simply presenting data, transforming it into a catalyst for positive change. The goal is not just to inform but to inspire action.

In addition to creating compelling reports, explore interactive dashboards. These dynamic tools empower users to explore data on their own terms, filtering and sorting information as needed. Interactive dashboards are particularly helpful when dealing with large datasets or when the audience has varying levels of expertise. They enhance user engagement and allow for deeper insights that static reports might miss. For example, if you're tracking many metrics, an interactive dashboard lets users focus on the most relevant ones to them.

Remember, the ultimate goal of reporting and visualization is to improve operational excellence. By presenting data clearly, concisely, and compellingly, you can empower your organization to make informed decisions, address bottlenecks, and drive positive change. Strong reporting doesn't just present facts; it fosters a culture of data-driven decision-making and contributes significantly to continuous improvement. It is, therefore, a critical component of achieving peak performance. Invest time in understanding your audience and the message you want to convey, and choose visual representations that best facilitate this communication. The effort will significantly pay off in enhancing your organization's ability to thrive.

4 Chapter 4: Risk Management

Risk is inherent in any project; ignoring it is a recipe for disaster. Understanding and proactively managing these potential problems is crucial for achieving operational excellence. This isn't about predicting the future flawlessly, but about building resilience and adaptability into your project's very core. It's about acknowledging that unexpected events will happen, and being prepared for them.

Identifying potential risks requires a systematic approach. Brainstorming sessions with your team are invaluable, drawing upon diverse perspectives and experience. Don't limit yourselves to obvious problems; consider unexpected delays in material delivery, sudden changes in market demand, and even unforeseen technological glitches. Consider using a risk register - a document that meticulously lists each identified risk, its potential impact, and its likelihood of occurrence. Categorize them, prioritize them, and then act.

Once you've identified the risks, analyze their potential impact. A simple high-low impact scale works wonders: High impact risks threaten the project's viability; low impact risks might cause minor delays or cost overruns. Then, estimate the likelihood of each risk occurring - again, use a simple scale, perhaps high, medium, and low probability. This framework allows you to prioritize risks based on a combination of impact and likelihood, focusing your mitigation efforts where they'll be most effective.

Mitigation strategies are the heart of effective risk management. This isn't about eliminating all risk—that's impossible. Instead, it's about reducing the likelihood of negative outcomes or minimizing their impact. For instance, if a high-impact risk is a potential supplier failure, mitigation might involve diversifying your suppliers, having a backup plan ready, or building in extra lead time. If a key employee leaving is a concern, then cross-training, succession planning, and robust documentation are crucial mitigation factors. Always plan for contingencies. Remember, risk management isn't a one-time event; it's an ongoing process. Regularly review your risk register, updating it as the project progresses and new information becomes available. Circumstances change; new risks emerge; old ones might diminish. Continuous monitoring enables you to adapt your strategies, adding new mitigations or adjusting existing ones as needed. Treat risk assessment as a dynamic component of your project's management, actively evolving alongside your progress. This adaptive approach ensures your project stays on track and that your operational excellence remains intact.

Flexibility and adaptability are key elements in navigating the uncertainties inherent in any project. Don't become wedded to inflexible plans. If a low-likelihood, high-impact risk materializes, don't stubbornly cling to your original strategy; instead, swiftly and decisively implement your pre-planned contingency measures. Avoid becoming paralyzed by fear of the unknown; rather, equip yourself with the knowledge and resources to react effectively when faced with unanticipated challenges. Efficient communication within the team is paramount.

Successful risk management requires clear communication across all levels of your team. Keep everyone informed of potential risks and the mitigation strategies in place. Transparency fosters a sense of shared responsibility and collective ownership of the project's success. Regularly update the team on risk assessment, sharing insights and analysis to ensure everyone is actively involved in managing potential problems. Open communication channels allow for rapid responses and collaborative problem-solving.

Investing in thorough risk management isn't just about avoiding problems; it's about creating a more robust and resilient project. By proactively identifying and mitigating potential risks, you build a more predictable and successful project path. Remember that careful planning, detailed analysis, and a dynamic risk management process are not mere additions to your operational excellence strategy; they are essential foundations. Your ability to manage risks will ultimately determine your ability to achieve your project goals efficiently and effectively, translating to sustainable long-term success. Proactive risk management contributes to increased efficiency, cost savings, and overall project success, making it a core tenet of operational excellence.

4.1 Identifying Potential Risks

Proactive risk identification is crucial for maintaining operational excellence. Ignoring potential problems can lead to significant setbacks, costing time, resources, and potentially damaging your reputation. Think of it like navigating a ship - a skilled captain doesn't simply set a course and hope for the best; they constantly monitor the horizon, anticipating potential storms and charting alternative routes.

A systematic approach is key. Begin by brainstorming potential risks across all aspects of your operations. Don't shy away from considering even seemingly minor issues; a small crack in the hull, if left unaddressed, can sink the entire ship. Engage your team in this process; they possess valuable on-the-ground insights often overlooked by upper management. Their perspective offers a richer, more comprehensive understanding of potential vulnerabilities.

Consider using a risk assessment matrix. This tool allows you to visually represent the likelihood and impact of each identified risk. Plot each risk on a grid, with likelihood on one axis and impact on the other. This will quickly highlight high-priority risks demanding immediate attention. Remember, a low-likelihood, high-impact risk, such as a major natural disaster, requires a contingency plan just as much as a high-likelihood, low-impact risk, like a minor equipment malfunction, needs preventative maintenance.

Once identified, analyze the root causes of each potential risk. Don't just treat the symptoms; dig deeper to understand the underlying issues. For instance, consistently late deliveries might stem from inadequate inventory management, unreliable suppliers, or poor communication. Addressing the root cause is far more effective than simply applying band-aid solutions. This deeper understanding will form the bedrock of your mitigation strategies.

Developing mitigation strategies involves creating plans to reduce the likelihood or impact of identified risks. For instance, a risk of supply chain disruption could be mitigated by diversifying suppliers, building up safety stock, or establishing strong relationships with key vendors. Similarly, risks related to employee turnover could be mitigated through improved employee engagement, competitive compensation packages, and professional development opportunities. These preventative measures are investments that protect your operational efficiency.

Effective communication is vital throughout the entire risk management process. Keep your team informed of identified risks and the mitigation strategies being implemented. Transparency builds trust and fosters a collaborative environment where everyone feels empowered to contribute to risk reduction. Regularly review and update your risk assessment, as circumstances change and new risks inevitably emerge. This ongoing process ensures your operations remain resilient and adaptable.

Remember, risk management isn't about eliminating all risks; that's impossible. It's about proactively identifying and mitigating potential threats, reducing their impact, and improving your organization's ability to weather unforeseen challenges. A well-executed risk management process contributes significantly to the overall operational excellence you're striving for. It's about building a more resilient, adaptable, and ultimately successful organization. The goal is not to be risk-free, but to be risk-aware and proactive in your response.

Don't underestimate the importance of regular reviews. Schedule recurring meetings to assess the effectiveness of your mitigation strategies. This ongoing evaluation ensures that your plans remain relevant and effective in the face of changing conditions. Continuously refine your approach, based on lessons learned and new information. Consider using key performance indicators (KPIs) to track the success of your risk mitigation efforts. This data-driven approach will provide valuable insights and guide future improvements.

Finally, remember that a successful risk management program is not a one-time event, but an ongoing, iterative process. The landscape of potential risks is constantly evolving. Regularly re-evaluate your assessments, refine your mitigation strategies, and communicate openly with your team. This commitment to continuous improvement is vital for sustaining operational excellence in the long term. A culture of proactive risk management contributes significantly to the overall resilience and success of your organization. By embracing this culture, you'll be well-equipped to navigate challenges and achieve sustainable operational excellence.

4.2 Mitigation Strategies

Proactive risk management isn't about predicting the future; it's about preparing for it. Understanding potential disruptions allows for strategic planning, minimizing their impact, and ultimately protecting your project's success. A robust risk management plan isn't a static document; it's a living, breathing strategy that evolves with your project.

Consider your project's unique challenges. What external factors could derail progress? Economic downturns, regulatory changes, and unexpected supply chain disruptions are all potential threats. Internally, insufficient resources, team conflicts, or a lack of clear communication can also significantly impact your outcomes. Identifying these potential issues early is paramount.

Once you've identified potential risks, analyze their likelihood and impact. A simple matrix can be helpful here, plotting the probability of each risk occurring against its potential negative consequences. This allows you to prioritize your efforts, focusing first on the most significant threats. High-probability, high-impact risks demand immediate attention; lower-probability, lower-impact risks can be addressed later.

Mitigation strategies aren't one-size-fits-all solutions. They need to be tailored to the specific risk you're facing. For example, if a major supplier's reliability is questionable, diversifying your supply chain is a crucial mitigation strategy. This reduces your dependence on a single vendor and protects against potential disruptions. Similarly, if you anticipate budget constraints, proactively securing additional funding or optimizing your budget is essential.

Effective communication is critical to successful risk mitigation. Keeping your team informed about potential threats and the strategies in place to address them fosters a sense of preparedness and encourages proactive problem-solving. Regular risk reviews, incorporating updates from team members, provide opportunities to adjust your strategy as the project progresses. Remember, unforeseen circumstances will always arise.

Don't underestimate the power of contingency planning. This involves developing alternative plans for handling various scenarios. What happens if a key team member leaves? What if a crucial piece of equipment fails? Having backup plans in place minimizes disruption and keeps your project on track. Contingency planning requires careful consideration and detailed planning, ensuring you have viable alternatives ready to implement should the need arise.

Building resilience into your project is another vital aspect of risk mitigation. This involves creating systems and processes that can withstand shocks and unexpected events. Having robust communication channels, clear roles and responsibilities, and a culture that values flexibility and adaptability are all key elements of resilience. A resilient project is better equipped to weather the storm and maintain its momentum even in the face of unforeseen challenges.

Finally, remember that risk mitigation is an iterative process. It requires constant monitoring, evaluation, and adjustment. Regularly review your risk register, updating it to reflect changes in the project landscape. Don't be afraid to adapt your mitigation strategies as new information emerges or circumstances evolve. A dynamic approach to risk management ensures your project remains resilient and on track, navigating unforeseen challenges effectively. Proactive engagement with risk management provides the essential foundation for sustained success.

5 Chapter 5: Continuous Improvement

Continuous improvement isn't a sprint; it's a marathon. Think of it as a journey, not a destination. You'll constantly be refining processes, adjusting strategies, and learning from both successes and setbacks. Embracing this mindset is crucial for sustained operational excellence. The key here isn't about radical overhauls, but small, incremental changes implemented consistently.

Kaizen, the Japanese philosophy of continuous improvement, offers a practical framework. It emphasizes small, incremental changes implemented consistently over time. Instead of aiming for huge leaps, focus on identifying small inefficiencies and systematically eliminating them. This approach reduces resistance to change, fostering a culture of ongoing improvement within your team. Remember, even minor adjustments can accumulate into significant improvements over time. This is about building momentum and a positive feedback loop of success.

Implementing change requires a structured approach. Start by clearly defining the problem you're addressing. Then, brainstorm potential solutions with your team. Involve everyone impacted by the change. Consider the root causes of the problem, not just the symptoms. This participatory process builds ownership and commitment, vital for successful implementation. This step is as important as implementing the changes themselves. Without it, you risk losing valuable insights and creating unnecessary hurdles for yourself and your team.

Monitoring progress is essential to assess the effectiveness of your changes. Establish key performance indicators (KPIs) to track your improvements. Regular monitoring allows for timely adjustments and course correction if needed. Don't just measure; analyze the data. Understand why you're seeing certain trends. This allows for a truly data-driven approach, optimizing your efforts and minimizing waste. This process is also critical in demonstrating the value of continuous improvement to stakeholders and securing continued support for your initiatives.

Let's consider a practical example. Imagine a customer service department experiencing high call wait times. A Kaizen approach would involve analyzing call data to identify bottlenecks, perhaps specific types of inquiries or time of day. Then, small changes could be implemented—additional training for agents handling complex issues, optimized call routing systems, or even improved self-service options on the company website. Tracking wait times, customer satisfaction scores, and agent handling times would provide clear evidence of the impact of these changes, allowing for further refinement.

Continuous improvement isn't solely about efficiency gains; it's about fostering a culture of learning and adaptation. Regularly review your processes. Seek feedback from your team, customers, and stakeholders. Encourage experimentation and innovation. Don't be afraid to try new things and learn from failures. This iterative process leads to ongoing refinement and breakthroughs. Remember that continuous learning and adaptation are essential for maintaining a competitive edge.

Moreover, the process must be integrated into the very fabric of your organization. Don't treat continuous improvement as a separate project but as an integral part of daily operations. Embed it into team meetings, performance reviews, and even casual conversations. Make it a natural part of your team's way of working. The goal is to create a self-correcting system where improvement is a continuous, ongoing effort. This integrated approach fosters a culture of innovation and agility, crucial for success in today's dynamic business environment.

The success of continuous improvement hinges on strong leadership. Leaders must champion the process, provide the necessary resources, and actively participate in improvement initiatives. They must also create a safe environment where team members feel comfortable sharing ideas, raising concerns, and suggesting improvements. Empowerment and trust are key to fostering a culture of continuous improvement within your teams. This leadership support is crucial for the long-term adoption of continuous improvement methodologies.

Remember, continuous improvement is a journey, not a destination. There will always be room for improvement, new challenges to overcome, and innovative solutions to implement. By embracing a Kaizen mindset, implementing changes strategically, and consistently monitoring progress, you'll drive significant improvements in your operations and achieve lasting operational excellence. The key is persistence and commitment to the process itself. Through continuous adaptation and a relentless focus on improvement, you can guide your team towards sustained success. Operational excellence isn't a one-time achievement; it's a continuous pursuit.

5.1 Kaizen Principles

Kaizen, the Japanese philosophy of continuous improvement, isn't just about incremental changes; it's a fundamental shift in mindset. It's about embedding a culture of constant refinement into the very fabric of your project, fostering a proactive approach to identifying and eliminating inefficiencies, big or small. This isn't a one-time fix, a quick solution; it's a journey, a commitment to ongoing progress. Think of it as the gentle, persistent erosion of obstacles, steadily shaping a more efficient, productive project landscape.

Implementing Kaizen effectively requires a systematic approach. Begin by identifying areas needing improvement. Don't overwhelm yourself; focus on one process, one bottleneck at a time. Engage your team; their insights are invaluable. Brainstorm potential solutions, however small they may seem. Remember, Kaizen values even the tiniest improvement as a step towards larger gains. Consider using tools like process mapping (detailed in Chapter 2) to visualize the workflow and identify choke points.

Once you've identified a target area, assemble a small, dedicated team to tackle it. Empower them with the authority to make decisions and implement changes. This promotes ownership and ensures swift action. Encourage open communication and collaboration. Everyone's ideas are important, fostering a culture where improvement suggestions are not only welcomed but actively sought. Remember the importance of effective communication, explored in detail in Chapter 6; clear and consistent communication is crucial for a successful Kaizen implementation.

The next stage is experimentation. Small-scale tests allow for quick feedback and adjustments. Don't be afraid to fail; experimentation often reveals unforeseen obstacles or opportunities. Each failure provides valuable learning, shaping the path towards a refined and optimal solution. Document the process carefully, recording successes, failures, and the rationale behind each decision. This meticulous record-keeping will form the foundation of your future improvement initiatives and provide valuable data for analysis, aligning with the data-driven decision-making strategies discussed in Chapter 3.

Once a successful improvement is identified, standardize it. Ensure the process is consistently implemented, minimizing the chances of regression. This standardization not only guarantees the perpetuation of your improvement but also creates a framework for applying the same principles to other areas. Remember to monitor progress closely, tracking key performance indicators (KPIs) to measure the effectiveness of your changes. The insights gained will inform further iterations, leading to a constant cycle of refinement.

The power of Kaizen lies in its simplicity and adaptability. It's not a rigid methodology; it's a flexible philosophy that can be adapted to any project, regardless of size or complexity. Embrace this adaptability; allow the process to evolve as your project and circumstances change. Always be on the lookout for opportunities to improve, and foster a culture of continuous improvement that permeates every aspect of the project. This proactive approach to improvement minimizes disruptions and enhances overall project performance.

Kaizen is not a solitary endeavor; it demands teamwork and collaboration. Each team member should be empowered to identify and suggest improvements, contributing to a shared objective. This collaborative spirit is essential, generating more creative solutions and fostering a sense of ownership. A culture of openness and mutual respect is crucial for successful Kaizen implementation. The collaborative environment discussed in Chapter 1 becomes paramount here.

The final key to effective Kaizen is persistence. Continuous improvement is an ongoing journey, not a destination. Don't expect immediate, dramatic results; small, consistent improvements add up over time. Celebrate even minor achievements; this positive reinforcement reinforces the Kaizen culture, encouraging continued engagement. The long-term strategies detailed in Chapter 7 build upon the foundation laid by consistent Kaizen implementation. Remember, the ultimate goal of Kaizen is not just operational excellence; it's the cultivation of a culture dedicated to continuous learning and refinement.

5.2 Implementing Change

Successful change implementation isn't about grand pronouncements; it's about meticulous planning and consistent execution. Begin by clearly articulating the 'why' behind the change. Employees need to understand the rationale, the benefits, and how it aligns with the overall organizational goals. Failure to clearly communicate this core purpose can lead to resistance and ultimately, project failure. This isn't just about sending out a memo; it requires open forums, Q&A sessions, and consistent reinforcement of the message at every level of the organization. Transparency and proactive communication are paramount.

Next, assemble a dedicated team committed to overseeing the change process. This team needs diverse skills and perspectives, representing all affected areas. Their role extends beyond mere implementation; they must anticipate potential roadblocks, adjust the approach as necessary, and consistently monitor progress against established milestones. Empower this team, giving them the authority to make decisions and resolve issues swiftly, avoiding lengthy bureaucratic processes. Remember, efficient decision-making is critical for a smooth transition.

Detailed planning is essential. Develop a comprehensive roadmap outlining the steps required, timelines for completion, and responsibilities for each team member. Break down large, complex changes into smaller, manageable tasks, allowing for more frequent progress assessments and fostering a sense of accomplishment along the way. This iterative approach keeps morale high and prevents the overwhelming feeling of a massive, insurmountable challenge. Regular updates are critical; these updates keep momentum and enthusiasm alive.

Training and development are not optional extras; they are fundamental to the successful implementation of change. Equip your team with the necessary knowledge and skills to operate within the new system or process. This includes hands-on training, mentorship, and readily accessible resources. The goal is to ensure everyone feels confident and competent, reducing anxiety and promoting buy-in. Don't underestimate the value of well-designed training programs; they significantly contribute to the success of the change initiative.

Throughout the implementation phase, actively monitor progress and collect feedback. Utilize key performance indicators (KPIs) to track whether the intended improvements are being realized. Regularly solicit feedback from team members, using various methods to encourage candid and honest responses. This feedback loop is crucial for identifying and addressing issues promptly, preventing them from escalating into larger problems. This iterative feedback system keeps the project on track and adaptable to any unforeseen challenges.

Finally, celebrate successes along the way. Acknowledge and reward individual and team contributions, reinforcing positive behavior and reinforcing the value of the change initiative. Recognition fosters a sense of ownership and engagement, encouraging continued commitment. It is essential to not only measure results but to also acknowledge and appreciate the effort that goes into achieving them; this keeps everyone motivated and encourages continued effort. The successful implementation of change is a team effort, and acknowledging that shared success is paramount for continued success. A final evaluation after the change will help understand what to improve in the future.

5.3 Monitoring Progress

Regular monitoring isn't just about checking boxes; it's about understanding the pulse of your project. Think of it as a conversation, a continuous dialogue between your plan and reality. Are you hitting your milestones? Are resources being used effectively? These aren't rhetorical questions; they're the heart of effective progress tracking. Consistent monitoring allows you to identify potential issues early, preventing them from snowballing into major problems later on. Early identification helps maintain control.

Effective monitoring requires more than just glancing at spreadsheets. You need a multifaceted approach, leveraging different methods to gain a comprehensive view. Regular meetings with the team provide invaluable qualitative feedback, capturing nuances that data alone might miss. Don't underestimate the power of direct observation; sometimes, seeing a process firsthand unveils hidden inefficiencies. These observations allow for real-time adjustments and fine-tuning.

Data analysis forms another crucial pillar. Refer back to Chapter 3; your carefully chosen KPIs, now analyzed, provide quantifiable evidence of progress—or the lack thereof. Compare your actual performance against your projected goals. Are you ahead of schedule? Falling behind? These answers dictate your next moves, allowing you to make data-driven decisions, ensuring your resources remain aligned with your objectives. Data can provide insight into potential issues, offering a proactive rather than a reactive approach.

Now, how do you translate this data into actionable insights? Don't get bogged down in endless spreadsheets. Visual representations—charts, graphs, dashboards—make complex data easily digestible. A well-designed dashboard provides a bird's-eye view of project health, highlighting key metrics at a glance. This enables rapid identification of problem areas, simplifying communication and speeding up responses. Consider creating a simple, yet effective, reporting system. This system shouldn't be overly complex, just concise and easy to understand. Regular reports, perhaps weekly or bi-weekly, should summarize progress against key goals, highlighting both successes and challenges. These reports should be concise, using visual elements to quickly convey the main points. Transparency is key; share these reports widely within the team to foster shared understanding and accountability.

Beyond the numbers, continuous feedback loops are essential. Establish regular feedback sessions with team members, soliciting their input on challenges and successes. Their on-the-ground perspective provides invaluable insights that might be missed by upper management. Encourage open communication; a culture of honest feedback is crucial for identifying and addressing issues proactively. This can lead to process improvements and better team morale.

Remember, monitoring progress isn't a one-size-fits-all endeavor. The optimal approach will vary depending on the project's complexity and goals. Experiment with different methods, adjusting your monitoring strategy based on what works best for your team and project. This iterative process ensures that your monitoring system continually evolves and adapts to your changing needs. Continuous refinement is a core component of operational excellence.

Regular reviews of your monitoring system itself are vital. Are the metrics you're tracking still relevant? Is the reporting frequency appropriate? Does your team find the system helpful and easy to use? Regular self-assessment ensures that your monitoring remains effective and efficient, preventing it from becoming a bureaucratic burden. A well-designed system should help, not hinder. Remember, the goal is not just to track progress, but to use that information to improve performance.

Finally, don't forget the human element. While data provides crucial insights, remember that your project is ultimately driven by people. Recognize and celebrate successes along the way. Acknowledge the team's hard work and dedication. This positive reinforcement strengthens morale and reinforces the commitment to continuous improvement. Celebrating milestones boosts team spirit and helps maintain momentum. Acknowledge the effort and dedication of the team; recognize their contributions. This fosters a positive and productive work environment, crucial for maintaining momentum and achieving long-term success. After all, successful projects are built not just on processes, but also on strong relationships and shared goals.

6 Chapter 6: Leading with Excellence

Empowering your team isn't simply about delegating tasks; it's about fostering a sense of ownership and responsibility. Give your employees the authority to make decisions within their areas of expertise. Trust their judgment, provide the necessary resources, and allow them the space to experiment and learn from their mistakes. This cultivates a more engaged and productive workforce, directly impacting the quality of their work and overall project success. Remember, empowering employees builds confidence and strengthens their commitment to the project's goals, leading to a more robust and self-sufficient team. Invest in their growth through training and mentorship programs; their success is directly tied to the project's triumph. Micromanagement stifles creativity and initiative. Empowering, on the other hand, unlocks potential and fosters a culture of innovation.

Collaboration isn't just a buzzword; it's the lifeblood of successful projects. Break down silos, encourage cross-functional communication, and create a collaborative environment where ideas flow freely. Implement tools and platforms that facilitate seamless information sharing and communication, perhaps even creating dedicated spaces for brainstorming and problem-solving. Foster a culture of open dialogue where everyone feels comfortable contributing their insights, regardless of their position or seniority. Regular team meetings, informal gatherings, and even social events can all help to build stronger relationships and improve collaboration. Remember that diverse perspectives enrich problem-solving, leading to more innovative and robust solutions.

Fostering innovation requires a deliberate shift in mindset. Encourage experimentation, tolerate calculated risks, and celebrate failures as learning opportunities. Create a safe space where team members feel comfortable proposing unconventional ideas without fear of judgment or retribution. Implement systems for capturing and evaluating innovative ideas, perhaps through suggestion boxes, dedicated brainstorming sessions, or regular idea pitches. Provide resources and support for developing and testing these ideas, allowing for iterative improvements and refinement. Reward innovative thinking and problem-solving; recognize and celebrate the successes of your team, both big and small. This positive reinforcement encourages a continuous cycle of creativity and improvement, transforming your team into a powerhouse of innovation.

Effective communication is the cornerstone of any successful project. Establish clear communication channels and protocols, ensuring everyone understands their responsibilities and how to access information. Regular updates, both formal and informal, are vital to keep everyone informed and aligned on the project's progress. Encourage open and honest feedback, both positive and constructive, to foster a culture of continuous improvement. Listen actively to your team members' concerns and suggestions, demonstrating that their voices are valued and their perspectives are important. Regular check-ins, both individual and group, help identify potential issues early, before they escalate into major problems. Remember that clear and consistent communication builds trust and fosters a collaborative atmosphere, essential for reaching peak performance.

Performance reviews shouldn't be dreaded events; they should be opportunities for growth and development. Focus on providing constructive feedback, both positive and negative, framed in a way that is supportive and motivating. Use data to support your assessment, highlighting both achievements and areas for improvement. Collaboratively create goals for future performance, setting realistic expectations and outlining a roadmap for success. Incorporate both quantitative and qualitative data, balancing objective measures of performance with subjective observations on behavior, teamwork, and problem-solving skills. Make it a two-way conversation; allow your team members to express their own perspectives, concerns, and aspirations. By making performance reviews a positive and constructive experience, you strengthen the employee-manager relationship, creating a culture of ongoing growth and mutual respect. Remember, your goal is not just to evaluate, but to empower and mentor, creating a dynamic environment for peak performance and continued excellence.

6.1 Empowering Employees

Empowering employees isn't about relinquishing control; it's about strategically distributing responsibility to unlock their potential and boost overall performance. Think of it as a carefully orchestrated symphony, where each instrument - each employee - plays a vital role in creating a harmonious whole. Effective empowerment involves trust, delegation, and a robust support system, creating an environment where individuals feel valued and capable of contributing meaningfully to the project's success. This isn't about throwing tasks over the wall and hoping for the best; it's about fostering ownership and accountability.

First, clearly define roles and responsibilities. Vague expectations breed confusion and frustration. Each team member needs a clear understanding of their contribution, their authority, and the expected outcomes. This clarity reduces ambiguity and empowers them to make informed decisions within their designated scope. Provide thorough training and resources, ensuring they possess the necessary skills and knowledge to succeed. A well-trained workforce is a confident workforce, more likely to embrace their empowered role.

Open communication is paramount. Create a culture where employees feel comfortable sharing ideas, concerns, and feedback without fear of reprisal. Regular check-ins, team meetings, and informal discussions foster open dialogue and ensure everyone feels heard and valued. This collaborative environment helps to identify and resolve potential issues proactively, preventing small problems from escalating into major roadblocks. Active listening, valuing diverse perspectives, and responding to concerns constructively are crucial components of this communication strategy. Remember that employees are not just cogs in a machine; they are intelligent, creative individuals who bring unique strengths to the table.

Next, foster a culture of trust. Trust is the bedrock of empowerment. When employees know that their leadership believes in their capabilities and trusts their judgment, they are more likely to take initiative, to innovate, and to contribute their best work. This isn't just about believing in their potential; it's about demonstrating that belief through actions. Delegating significant tasks, providing autonomy in decision-making, and offering constructive feedback rather than criticism all convey trust and demonstrate that leadership values their contributions. Micro-managing undermines empowerment and creates an atmosphere of distrust.

Furthermore, invest in your employees' growth and development. Empowerment goes beyond simply assigning tasks; it's about nurturing potential. Provide opportunities for professional development, such as workshops, conferences, or mentorship programs. Support their pursuit of new skills and knowledge, not only benefiting the individual but also enhancing their value to the project. Investing in your team's growth signals your commitment to their success and encourages them to invest their efforts and dedication in return. This cyclical investment fosters loyalty and a sense of shared growth.

Finally, recognize and reward achievements. Acknowledging contributions - both big and small - fosters a sense of accomplishment and reinforces positive behavior. Public recognition, bonuses, or simple words of appreciation all contribute to a culture of appreciation. Celebrating milestones boosts morale and reinforces the importance of individual and team contributions. This positive reinforcement encourages continued effort and reinforces the belief that their contributions are valued and their hard work is appreciated. Remember, a motivated and empowered workforce is a highly productive workforce. By implementing these strategies, you're not just empowering employees; you're unlocking the full potential of your team, driving project success, and fostering a thriving work environment.

6.2 Promoting Collaboration

Effective collaboration isn't merely about people working together; it's about creating a synergistic environment where the collective output surpasses the sum of individual contributions. This requires a deliberate and multifaceted approach, moving beyond simple teamwork to true collaboration. It necessitates a conscious effort to build trust, foster open communication, and establish shared goals. Think of it as constructing a finely tuned orchestra, where each instrument, while distinct, plays a vital role in the overall harmony.

Open communication is the bedrock of successful collaboration. This doesn't just mean frequent meetings; it requires active listening, transparent information sharing, and a culture where individuals feel comfortable voicing their opinions, even if they differ from the majority. Establish clear communication channels, utilizing both formal and informal methods to ensure information flows efficiently and effectively. Regular check-ins, both individual and group, help maintain momentum and address potential roadblocks early on. Remember, a well-informed team is a powerful team.

Building trust is an iterative process that demands consistent effort and genuine engagement. Start by fostering a respectful and inclusive environment where everyone feels valued and heard. Encourage open dialogue, acknowledge contributions, both large and small, and actively address conflicts fairly and promptly. Demonstrate vulnerability by sharing your own challenges and uncertainties; this builds authenticity and creates a safe space for others to do the same. Remember, trust isn't given; it's earned over time through consistent actions.

Shared goals are the compass guiding collaborative efforts. Clearly defining objectives and ensuring everyone understands their roles and responsibilities within the overall plan is crucial. This involves not simply assigning tasks but fostering a collective ownership of the project's success. Regular progress reviews, where the team collectively assesses achievements and identifies challenges, are essential for staying on track and reinforcing shared commitment. Everyone needs to understand how their work connects to the bigger picture.

Leveraging diverse perspectives is key to innovative solutions. Encourage team members to challenge assumptions and offer alternative viewpoints. Embrace healthy debate and disagreement as opportunities for learning and growth. Different backgrounds and experiences bring unique strengths to the table, leading to more comprehensive and creative outcomes. Structured brainstorming sessions, with clearly defined rules of engagement, can further stimulate diverse thinking and idea generation. The power of collaboration lies in its ability to surpass individual limitations.

Conflict resolution is an inevitable part of any collaborative endeavor. When conflicts arise, address them promptly and constructively. Focus on understanding the underlying issues rather than assigning blame. Facilitate open dialogue, encouraging all parties to express their perspectives. Seek mutually agreeable solutions, ensuring that everyone feels their concerns have been heard and addressed. Remember, managed conflict can strengthen relationships and foster deeper understanding. It is a catalyst for progress.

Tools and technologies can greatly enhance collaboration, provided they are chosen and implemented strategically. Choose tools that facilitate communication, information sharing, and project management. Ensure that all team members are proficient in using the chosen technologies and provide adequate training if necessary. Consider the team's size, geographical distribution, and project complexity when selecting tools. The right technology can streamline workflows and enhance overall efficiency but only if properly implemented.

Finally, celebrate successes, both large and small, to foster a sense of collective achievement and reinforce positive behaviors. Recognition and appreciation boost morale and encourage continued collaboration. Consider using both formal and informal methods of recognition, tailoring the approach to the team's culture and preferences. Remember, celebrating accomplishments reinforces positive behaviors and builds a stronger, more cohesive team. The journey is as important as the destination. Sustained collaboration builds not just successful projects but also strong, enduring teams.

6.3 Fostering Innovation

Cultivating a culture of innovation isn't merely about brainstorming sessions; it's about fundamentally changing how your team operates. It requires a deliberate and sustained effort, woven into the fabric of your daily routines. Think of it as building fertile ground for new ideas to sprout and flourish, not a one-time event.

Start by empowering your team. Provide them with the autonomy to experiment, to fail, and to learn from their mistakes without fear of retribution. This doesn't mean abandoning accountability; it means fostering a supportive environment where calculated risks are encouraged. Frame failures not as defeats, but as valuable learning experiences, paving the way for future successes. Remember, some of history's greatest innovations emerged from initial failures.

Encourage cross-functional collaboration. Break down the silos that often stifle creativity. By bringing together individuals from different departments, you create a melting pot of perspectives and expertise. This intermingling of ideas sparks unexpected connections and innovations that wouldn't emerge in isolated teams. Organize workshops, team-building activities, and cross-departmental projects specifically designed to foster this collaboration. The resulting synergy can dramatically boost your organization's innovative capacity.

Regularly dedicate time for dedicated brainstorming sessions. Structure these sessions effectively to maximize their impact. Employ techniques like "brainwriting," where participants write down ideas individually before sharing, preventing the dominance of vocal individuals. Use visual aids, mind maps, and other tools to help generate and organize ideas. Establish clear goals for each session to keep the focus sharp and prevent unproductive tangents. Remember, the goal isn't just quantity, but quality of ideas, and a structured approach helps achieve that.

Invest in training and development. Equip your team with the skills and knowledge they need to innovate. This goes beyond technical skills; it includes creativity workshops, design thinking training, and sessions on future trends relevant to your industry. A well-trained workforce is a more adaptable and innovative workforce. Consider external training programs, mentorship schemes and internal knowledge sharing to achieve a holistic approach to upskilling.

Implement a robust idea management system. Create a formal process for capturing, evaluating, and implementing new ideas. This system should be easily accessible, transparent, and encourage participation. Ensure there is a clear process for feedback, enabling employees to understand why certain ideas are chosen or rejected and ensuring that rejected ideas aren't dismissed but rather are constructively critiqued. The more inclusive your idea management system is, the richer the pool of creative ideas will be.

Embrace experimentation and rapid prototyping. Don't be afraid to try new things, even if they seem risky. Encourage quick, iterative cycles of development and testing to validate ideas and adapt them as needed. Lean methodologies such as agile development can be instrumental in facilitating this process. By constantly testing and refining, you minimize wasted effort and maximize the chances of developing successful innovations.

Recognize and reward innovation. Celebrate successes, both big and small. Acknowledge individuals and teams who contribute to innovation, reinforcing a positive feedback loop. Financial rewards are useful, but non-financial recognition—public acknowledgement, awards, or opportunities for increased responsibility—can be just as effective in motivating future innovation. Make sure your reward system clearly reflects the importance you place on innovation. It must be something more than a token gesture; it should feel meaningful to the recipients.

Remember, fostering innovation is a continuous journey, not a destination. It requires consistent effort, commitment, and adaptation. By embedding these principles into your organizational culture, you'll not only generate a steady stream of innovative ideas but also foster a more engaged, motivated, and resilient workforce capable of navigating the challenges and opportunities of a rapidly changing world. The key is to actively cultivate an environment where creativity thrives and innovation is expected and rewarded. Doing so will allow you to secure your company's competitive edge in an ever-evolving marketplace. Your ongoing commitment to these principles will yield substantial returns, ensuring sustainable growth and future success. The journey to fostering innovation is a long-term investment which, once embraced holistically, becomes a valuable asset for your business. The efforts you put in will be well worth the reward, offering lasting benefits to your organization.

6.4 Effective Communication

Effective communication isn't just about speaking; it's about understanding. Truly effective communication within a project necessitates a multifaceted approach, going beyond simple information exchange to foster genuine collaboration and shared understanding. It's the cornerstone of a high-performing team, ensuring everyone is aligned on goals, processes, and expectations. Without clear, consistent communication, even the most meticulously planned project risks derailing due to misunderstandings, missed deadlines, and duplicated efforts.

Building a strong communication foundation starts with identifying your audience. Are you addressing a team of seasoned professionals, junior members requiring more detailed explanations, or external stakeholders needing concise updates? Tailoring your message to each audience's understanding and needs is crucial for effective delivery. Avoid jargon or overly technical terms unless everyone is familiar with them. Remember, clarity trumps complexity. A simple, direct message, delivered in the appropriate format, resonates far more powerfully than a convoluted, jargon-filled presentation.

Next, choose the right communication channel. A quick question might be best addressed through instant messaging, while complex project updates often require a formal meeting or email. Consider the urgency, sensitivity, and complexity of the information when selecting your preferred method. While email remains a useful tool, face-to-face communication fosters stronger relationships and allows for immediate clarification of any ambiguities. Consider using visual aids like diagrams and charts to convey information more effectively. This visual reinforcement can significantly improve understanding and retention, particularly for complex information. Don't underestimate the power of a well-designed infographic.

Active listening is as important as clear speaking. Truly hearing your team members' concerns, suggestions, and feedback demonstrates respect and fosters trust. Creating a safe space for open dialogue, free from judgment or interruption, is key to constructive communication. Encourage questions and feedback at every stage of the project. This open exchange of ideas not only prevents misunderstandings but also taps into the collective intelligence of your team, fostering innovation and problem-solving. Remember to actively listen and respond appropriately, reflecting what you've heard before answering to ensure you've truly understood.

Regular feedback mechanisms are crucial. This includes regular team meetings, one-on-one check-ins, and project milestone reviews. These opportunities allow for early detection of potential communication breakdowns and immediate corrective actions. Don't wait until a problem becomes critical before addressing it; proactive communication is essential. Schedule regular meetings, utilize project management software for updates, and ensure there is always an accessible point of contact for questions or concerns. This provides transparency and promotes a sense of collective responsibility.

Finally, remember that effective communication is an ongoing process. Continuous improvement means regularly evaluating your communication strategies and adapting them to ensure they are meeting the project's evolving needs. Feedback from team members is invaluable in this process. Gather feedback through informal channels, formal surveys, or dedicated reviews, utilizing the information obtained to refine your communication processes and enhance overall team performance. This commitment to ongoing improvement ensures your communication stays aligned with the project's dynamic nature and fosters a collaborative, productive environment.

6.5 Performance Reviews

Performance reviews are more than just annual check-ins; they're vital instruments for fostering growth and aligning individual contributions with organizational objectives. A poorly executed review can damage morale and stifle productivity, while a well-structured one can ignite passion and drive exceptional performance. Therefore, approaching performance reviews with meticulous planning and a genuine commitment to employee development is paramount.

To begin, ensure your performance review process is meticulously documented and consistently applied across the board. This consistency prevents bias and ensures fairness, building trust and confidence among your team. Transparency is key; let your employees know exactly what criteria will be assessed, how their performance will be measured, and what the potential outcomes of the review are. This proactive approach eliminates ambiguity and sets clear expectations.

Preparation is absolutely crucial. Gather data well in advance. Don't rely solely on your memory; document key achievements, challenges faced, and areas for improvement throughout the year. This data should include specific examples of both successful projects and instances where improvement is needed. Consider using a 360-degree feedback approach where possible, gathering input from peers, subordinates, and clients, offering a more holistic view of the employee's performance. This comprehensive approach provides valuable perspectives often overlooked in a traditional manager-employee dynamic.

Conducting the review itself demands careful consideration. Create a comfortable and safe space for open dialogue. Begin by acknowledging the employee's contributions and celebrating successes. Positive reinforcement sets a constructive tone and makes the entire process feel less like a judgment and more like a collaborative discussion about growth. Next, move into the constructive feedback, focusing on both strengths and areas for improvement. Always frame the feedback in terms of behavior and impact, avoiding vague generalizations or personality attacks. Remember to provide specific, actionable steps for improvement. The employee shouldn't leave the meeting feeling overwhelmed or demoralized, instead, they should leave feeling empowered to enhance their skills and contributions.

The post-review phase is just as critical. Ensure that the employee understands the expectations for improvement and has a clear action plan in place. This could include setting SMART goals—Specific, Measurable, Achievable, Relevant, and Time-bound—to track progress. Schedule follow-up meetings to review progress and provide ongoing support. Remember, the performance review isn't a one-time event; it's the start of an ongoing commitment to employee development and high-performance. Document the entire process, including goals, agreed upon action steps, and dates for follow-up. This documentation protects both the employee and the organization.

Finally, integrate the lessons learned from these reviews into your ongoing management style. Regularly assess your methods and refine them for greater effectiveness. Pay attention to which strategies work best and adjust your approach as necessary. Analyze the feedback you receive from employees and use that feedback to strengthen your leadership skills and improve the overall performance review process. Remember, fostering a culture where performance reviews are perceived as opportunities for growth and development is vital for ensuring sustained high performance within your organization. Continuous improvement applies to your leadership style as well, and the performance review process is an excellent barometer to measure that success. By consistently refining your approach, you cultivate a culture of continuous improvement, ensuring that every employee is empowered to reach their full potential and, in turn, contribute to the company's overall success. Regularly evaluating the effectiveness of your performance review system ensures you remain responsive to evolving needs and maintain a truly beneficial process for everyone involved. Continuous evaluation and adaptation of this process reflect your commitment to employee growth and the overarching success of your team, further strengthening the organizational alignment and operational excellence you strive to achieve.

7 Chapter 7: Sustaining Success

Maintaining operational excellence isn't a destination; it's a continuous journey. Success demands a proactive, adaptable approach, far beyond the initial implementation of strategies detailed in previous chapters. Ignoring this crucial aspect risks regression, undermining all previous efforts. Sustaining triumph requires a long-term vision, a willingness to evolve, and a commitment to continuous learning.

Building upon the foundations established thus far, we now focus on enduring strategies. This means shifting from project-based thinking to a systemic approach. We're not just fixing immediate problems; we're creating resilient processes capable of withstanding future challenges. This requires robust systems and clear escalation paths, ensuring issues don't spiral into major crises. Regular reviews, perhaps quarterly or even monthly, analyzing performance against targets and identifying potential threats, become essential.

A critical element in sustaining success is embracing change. The business landscape is constantly shifting. New technologies emerge, market demands fluctuate, and unforeseen events disrupt even the best-laid plans. Rigidity is the enemy of long-term success. Therefore, fostering a culture of adaptability, where continuous learning and experimentation are valued, is paramount. This doesn't mean abandoning your core principles; rather, it signifies a willingness to refine and adjust strategies based on real-time feedback and emerging insights. Develop clear channels for employees to voice concerns and suggest improvements; actively solicit feedback and incorporate it into your processes.

Regularly review and update your process maps. What worked flawlessly six months ago might now present bottlenecks. Identify those friction points and implement solutions before they severely impact efficiency. Regularly auditing your systems and procedures ensures that the improvements you've meticulously implemented remain effective and aren't slowly eroded by complacency. The key is not just reacting to problems, but proactively anticipating potential issues and mitigating them before they arise.

Moreover, fostering a culture of continuous improvement is not merely a buzzword; it's the lifeblood of sustained success. Regular training and development programs for your team should be a priority. This isn't just about technical skills; it's about equipping employees with the analytical and problem-solving skills needed to adapt to new challenges and propose creative solutions. Empower your teams to suggest improvements, and actively encourage experimentation and innovation. Recognize and reward individuals who identify and resolve problems, fostering a sense of ownership and commitment.

Finally, remember that sustained success hinges upon consistent communication and open feedback loops. Keep all stakeholders informed about progress, challenges, and future plans. This includes regular updates to senior management, transparent communication with teams, and open dialogues with clients. Transparency builds trust and buy-in, crucial for long-term commitment to the operational excellence model. Celebrate successes and openly discuss challenges; use setbacks as learning opportunities to strengthen processes and prevent similar incidents in the future. Building this culture ensures that everyone understands the importance of sustaining operational excellence, actively contributing to its ongoing success. The pursuit of excellence is a marathon, not a sprint; consistent effort and a commitment to continuous improvement are the keys to long-term triumph.

7.1 Long-Term Strategies

Cultivating long-term success in operational excellence requires a proactive, multifaceted approach that extends beyond immediate improvements. It's not just about achieving peak performance today; it's about building a resilient system capable of weathering future storms and consistently delivering exceptional results. This involves strategic foresight, continuous adaptation, and a commitment to ongoing learning and development.

A crucial element is the creation of a flexible organizational structure. Rigid hierarchies can hinder adaptation to change. Instead, consider fostering a flatter organizational structure that encourages collaboration and empowers employees at all levels to contribute to strategic decision-making. This decentralized approach allows for faster responses to emerging challenges and opportunities, enabling the organization to stay agile and competitive. Empowerment translates to increased ownership and commitment, resulting in more effective problem-solving and innovation.

Investing in employee development is paramount. This isn't solely about providing technical training, although that's certainly important. It's about cultivating a culture of continuous learning where employees are encouraged to acquire new skills, explore innovative approaches, and develop their leadership potential. This commitment to growth will not only enhance individual performance but also create a dynamic, forward-thinking workforce capable of navigating the complexities of a rapidly changing business landscape. Consider implementing mentorship programs, providing access to online learning resources, and supporting employees' participation in industry conferences and workshops. The return on investment in human capital is significant, ensuring a sustainable competitive edge.

Establishing robust systems and processes is another cornerstone of long-term success. While previous chapters discussed process mapping and optimization, maintaining these systems requires ongoing monitoring and refinement. Regular audits should be conducted to identify areas for improvement, ensuring that processes remain efficient and effective. This systematic approach prevents gradual degradation and maintains a high standard of operational excellence. Invest in technology that supports streamlined workflows, automates repetitive tasks, and provides real-time data analysis. This ensures efficient resource allocation and enables prompt responses to changing market demands.

Building strong relationships with stakeholders is crucial for sustained success. This includes maintaining open communication with customers, suppliers, and partners, actively seeking their feedback and incorporating it into strategic planning. Long-term success relies on mutual benefit and collaboration. By fostering trust and understanding, you build strong, resilient partnerships that can withstand market fluctuations and competitive pressures. A strong network of stakeholders can provide invaluable support and insights, contributing significantly to long-term stability and growth.

Finally, establishing a culture of continuous improvement is essential. This requires a commitment from leadership to champion change, actively encourage experimentation, and celebrate successes. Embrace a culture of learning from both successes and failures, fostering a mindset where continuous improvement is viewed not as a burden but as an ongoing journey of growth and enhancement. This will ensure that the organization remains adaptable and resilient, capable of thriving in an ever-changing environment. This ingrained culture ensures that the pursuit of operational excellence is not a temporary project but a defining characteristic of the organization's identity. It creates a sustainable system where improvement is a continuous process, ensuring enduring success.

7.2 Adapting to Change

Change is inevitable. It's a constant, a force shaping industries, markets, and even the very fabric of our daily operations. Ignoring this reality is a recipe for stagnation, a slow descent into irrelevance. Embracing change, however, is the key to not just survival, but to thriving in a dynamic environment. This isn't about reacting to shifts; it's about proactively anticipating them and integrating adaptability into the very core of your operational strategy.

Successful adaptation doesn't happen overnight. It's a gradual process that necessitates a shift in mindset. It starts with fostering a culture of learning and curiosity within your organization. This means encouraging employees to embrace new ideas, to experiment, and even to fail - for failure provides invaluable lessons. Regular training, both technical and soft skills-focused, is also crucial in equipping your workforce to navigate the evolving landscape. Encourage knowledge sharing across teams; build in time for brainstorming and collaborative problem-solving. These are the building blocks of a resilient, adaptable operation.

Develop robust, flexible systems and processes. Rigid, inflexible structures are easily broken by unexpected changes. Instead, design your operational framework with adaptability in mind. Implement modular processes that can be adjusted or reconfigured as needed without disrupting the entire system. Consider incorporating flexible resource allocation strategies that allow for quick reallocation of personnel and materials based on changing priorities. Investing in scalable technology that can adapt to increasing demands is also a critical step.

Monitor the external environment continuously. Stay abreast of industry trends, technological advancements, and shifts in consumer demand. Use market research, competitive analysis, and industry publications to inform your strategic planning. This isn't a passive observation; it's about proactively identifying potential disruptions and adjusting your strategy accordingly. Regularly review your key performance indicators (KPIs) to identify early warning signs of change and take preemptive measures to mitigate potential risks. This proactive approach allows for a smoother transition and reduces the likelihood of drastic, disruptive adjustments.

Leadership plays a pivotal role in driving adaptation. Leaders must not only model adaptability themselves but must also cultivate a culture that embraces it. This means fostering open communication, providing transparent updates on changes, and encouraging feedback from all levels of the organization. Empower your employees to propose solutions, and recognize and reward initiative. Create a safe environment where employees feel comfortable suggesting improvements and experimenting with new approaches without fear of retribution. Leadership's role is to be a guiding force, steering the organization through change while keeping the team motivated and focused.

Beyond simply reacting to change, proactive anticipation offers immense advantages. By analyzing trends and anticipating future demands, you can position your organization to capitalize on emerging opportunities. This might involve investing in new technologies before your competitors, developing new products or services to meet evolving needs, or expanding into new markets before they become saturated. This proactive stance not only mitigates risks but also creates opportunities for growth and innovation, transforming adaptation into a strategic advantage.

Remember, adapting to change is a continuous journey, not a destination. It requires ongoing evaluation, refinement, and a commitment to lifelong learning. Regularly review your adaptation strategies, assess their effectiveness, and make necessary adjustments. Embrace feedback, both internal and external, and use it to refine your approach. This continuous improvement cycle is essential for maintaining agility and ensuring long-term success in a constantly evolving world. By incorporating these strategies, you'll not only weather the storms of change but harness its power to drive your organization toward unparalleled success.

www.ingramcontent.com/pod-product-compliance
Lightning Source LLC
LaVergne TN
LVHW080531210926
848481LV00021B/2551
* 9 7 9 8 3 4 9 2 8 3 3 9 0 *